BEI GRIN MACHT SICH IHR WISSEN BEZAHLT

- Wir veröffentlichen Ihre Hausarbeit, Bachelor- und Masterarbeit

- Ihr eigenes eBook und Buch - weltweit in allen wichtigen Shops

- Verdienen Sie an jedem Verkauf

Jetzt bei www.GRIN.com hochladen und kostenlos publizieren

Impressum:

Copyright © 2018 GRIN Verlag
Druck und Bindung: Books on Demand GmbH, Norderstedt Germany
ISBN: 9783668780477

Suresh Aluvihara

Study of the Fundamental Corrosion of Ferrous Metals in Crude Oils

GRIN Verlag

Study of the Fundamental Corrosion of Ferrous Metals in Crude Oils

S. Aluvihara

[1]*Department of Science and Technology, Uva Wellassa University, Badulla, Sri Lanka.*

Abstract

Corrosion is a crowing term regarding the ferrous metals which accomplishes premonitory issues on such metals with backing of the conditions of surrounding environment. According to the consistency ofcrude oils it tends to bear a hand of metallic corrosion foremost the effect of sulfur compounds, salts and organic acids that presence in crude oils. The investigation of the corrosion rates of seven different types of ferrous metals referring to both Murban and Das Blend crude oils and some essential episodes regarding the corrosion were the leading intention of the research. The sulfur contents, salt contents, mercaptans contents and acidities of both crude oils were determined by in succeed of XRF analyzer, salt analyzer and titration methods while testing the chemical composition of each metal by XRF detector. A batch of equal sized metal coupons was prepared from each type of metal and dipped entirely in both crude oils separately. The rate of corrosion in each metal coupon was determined by the weight loss method after 15, 30 and 45 days from the immersion while observing the corroded surface of each coupon by an optical microscope. The ferrous concentrations and copper concentrations in crude oil samples weretested by the AAS after the interactions with metals and the hardness of each metal coupon was tested before the immersion in crude oil and after corroded. There were found relatively lower corrosion rates from stainless steels than other the metals while finding some higher metallic concentrations in crude oils regarding carbon steels, slight reduction of the hardness after the corrosion and finally it can be concluded the formation of FeS, corrosion cracks and pitting corrosion in most of occasions with the microscopic observations regarding the appearances of such a compound.

Keywords: Ferrous metals, Crude oils, Corrosion rate, Corrosion compounds, Hardness

Content

Introduction

Corrosion is a common term that ever heard regarding the ferrous metals. Usually it is defined as a result of the interaction between the metal and the surrounding medium of that relevant metal. In the industry of crude oil refining the interaction between crude oils and ferrous metals can be emphasized as the key phase of the metallic corrosion [1] [3]. The contents of sulfur, active sulfur compounds, organic acids and salts in crude oils play the major task in the cause of metallic corrosion [2] [4]. According the terms of active sulfur compounds mercaptans are the foremost corrosive compound with the contribution of the essential conditions of the environmental such as the temperature. The chemical composition of such metal also reserved some significant consideration regarding the corrosion. In most of occasions the corrosion phenomenon is found near the heat exchangers, column head, transportation tubes and raw crude storage tanks. In the current research there were expected to study the nature of corrosion of ferrous metals in both Murban and Das Blend crude oils, identify the most suitable ferrous metal for the usages of the industry of crude oil refining and investigate the variations of the properties of such metals due the corrosion such as the hardness. Also it was expected to speculate the strength of the corrosive properties of both Murban and Das Blend crude oils in the cause of corrosion in seven different types of ferrous metals.

Materials and Methodology

System Modeling

Metals

The metal sample was consisted three different types of carbon steels, three different types of stainless steels and Monel which has trace amount of ferrous that every metal is used in the different sections regarding the crude oil refining process.

Crude Oils

Murban and Das Blend were selected as the samples of crude oils. Crude oil is a mixture of hydrocarbons. Those are different exiguously in their chemical composition including corrosive compounds according to their geological formation. Murban and Das Blend are the two types of crude oils which are meager difference in their chemical compositions and also recently used in the crude oil refining industry of Sri Lanka.

Methodology

Corrosive Properties of Crude Oils

In this research four predominant corrosive properties and their effects were considered and tested as the methods given in the Table 1.

Table 1: Determination of the corrosive properties of crude oils

Property	Method	Readings
Sulfur content	Directly used.	Direct reading
Acidity	Each sample was dissolved in a mixture of toluene and isopropyl and titrated with potassium hydroxide.	End point
Mercaptans content	Each sample was dissolved in sodium acetate and titrated with silver nitrate.	End point
Salt content	Each sample was dissolved in organic solvent and exposed to the cell of analyzer.	Direct reading

Chemical Compositions of Metals

The elemental compositions metal samples from each type of metal were determined by the XRF detector. According to the sensitivity of XRF detector it could be determined the percentage of every metallic elements and most of non-metallic elements exclusive of carbon in each particular metal.

Determination of the Corrosion Rates of Metals and Weight Loss Method

A batch of similar sized metal coupons was prepared from seven different types of ferrous metals and cleaned the surface of each metal coupon until devoid of any extrinsic compound on the surface of each metal coupon. The metal coupons were immersed in both crude oils separately and homogeneously in the room temperature. The setup of apparatus is given in the Figure 1.

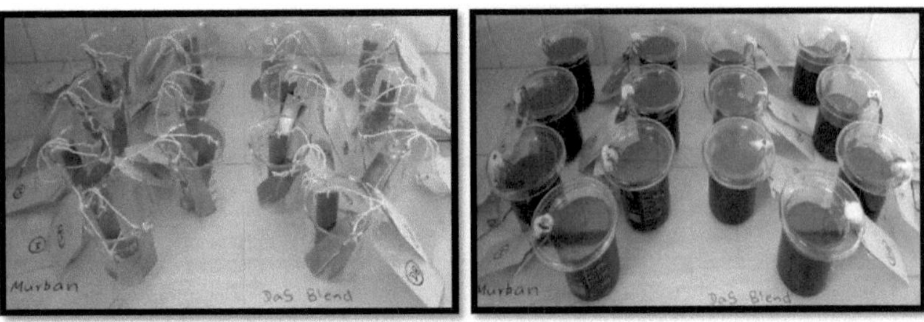

Figure 1: (a) Samples and (b) apparatus setup

The corrosion rate of each type of metal was detremined in order to after 15, 30 and 45 days from the immersion with respect to the both crude oils by the weight loss method [10]. The corroded metal surfaces were cleaned by sand papers and isooctane while observing the surfaces through the 400X lens of an optical microscope.The mathematical expression and terms are defined in the Equation 1.

$$CR = \frac{W}{(D \times A \times t)} \times k$$

.............(Eq.1)

Where;

W = weight loss in grams

k = constant (22,300)

D = metal density in g/cm^3

A = area of metal piece ($inch^2$)

t = time (days)

CR= Corrosion rate of metal piece

Qualitative Analysis of the Corrosion

According to the qualitative analysis the surfaces of metals were observed through the 400X lens of an optical microscope before the immersion in crude oils and after the formation of corrosion on the surface. The identification of corrosion compounds with the aid of their visible appearances and confirmation of the corrosion compound were the foremost objectives regarding the qualitative analysis.

Interaction of Metals and Crude Oils

According to the interaction between metals and crude oils there may be a tendency to decay some metallic compounds into the surrounding medium. In the current research it was expected to test the concentrations of ferrous and copper in crude oil samples after the interaction with metals as a confirmation stage for the formation of corrosion on metals. In the methodology each 1 ml of each crude oil sample was diluted with 9 ml of 2-propanol and the precipitations were filtered out.

Effect of Corrosion on the Properties of Metals

Corrosion is a phenomenon which is occurred on the surface of relevant metal. Therefore it can be directly affected on the conditions of the properties on the surface such as the hardness [3]. The initial hardness and the hardness after corroded were tested in each metal coupon based on the ambition of investigating the deviation of such a property. The instrument was the Vicker's hardness tester which gives a certain numerical value for the hardness regarding the particular point. There were interpreted an average value for the hardness of each metal coupon after testing at least three random points on each metal coupon.

Results and Discussion

Corrosive Properties of Crude Oils

According to the observed results the values for the particular corrosive properties of crude oils are given in the Table 2.

Table 2: Corrosive properties of crude oils

Property	Murban	Das Blend
Sulfur content (Wt. %)	0.758	1.135
Salt content (ptb)	4.4	3.6
Acidity (mg KOH/g)	0.01	0.02
Mercaptans content (ppm)	25	56

When comparing the contents of elemental sulfur and active sulfur in both crude oils Das Blend was composed some greater amounts in those two compounds than the

Murban crude oil. Sulfur and active sulfur compounds having a fraction or functional group that can be reacted with metals with the aid of water and caused the metallic corrosion which is known as the process "sulfidation" also highly depended on the temperature typically at 230^0 C [2] [6]. Mercaptans also an active sulfur compound which is having a general formula of RSH. The general chemical reaction of sulfidation is given in the Equation 2.

$$8 \, Fe + S_8 \longrightarrow 8 \, FeS \ldots\ldots\ldots\ldots (Eq.2)$$

According to the salt contents in both crude oils it was found relatively higher amount of salts in Murban crude oil than Das Blend crude oil. The content of salts in crude oils is including the summation of $CaCl_2$, $MgCl_2$ and $NaCl$. Due to the chemical reaction between water and salts at some higher temperatures those salts tend to be broken into HCl and later those HCl tend to behave as highly corrosive compounds when the system is approaching to low temperatures due to the reaction between HCl molecules and moisture to produce hydrochloric acids as a mechanism of reproducing HCl [7]. Usually it tends to form the metallic sulfides on the metal surface itself as discussed under the chemical reactions in Equation 3, Equation 4 and Equation 5.

$$CaCl_2 + H_2O \longrightarrow CaO + 2HCl \ldots\ldots\ldots\ldots (Eq.3)$$
$$HCl + Fe \longrightarrow FeCl_2 + H_2 \ldots\ldots\ldots\ldots\ldots (Eq.4)$$
$$FeCl_2 + H_2S \longrightarrow FeS + 2HCl \ldots\ldots\ldots\ldots (Eq.5)$$

According to the compositions of organic acids in both crude oils Das Blend crude oil was contained higher amount of organic acids than the Murban crude oil. According to the geological formation of crude oils they might have composed with some significant amount of organic acids also known as naphthenic acids which are having the general formula of RCOOH [2] [4] [9]. The reaction between those organic acids and ferrous metals are explained in the Equation 6, Equation 7 and Equation 8.

$$Fe + 2 \, RCOOH \longrightarrow Fe \, (RCOO)_2 + H_2 \ldots\ldots\ldots\ldots (Eq.6)$$
$$FeS + 2 \, RCOOH \longrightarrow Fe \, (COOR)_2 + H_2S \ldots\ldots\ldots\ldots (Eq.7)$$
$$Fe(COOR)_2 + H_2S \longrightarrow FeS + 2 \, RCOOH \ldots\ldots\ldots\ldots (Eq. 8)$$

FeS is the major corrosion compound that formed as the result of above chemical reactions. When considering the overall effect of those corrosive properties Das Blend sowed some stronger effect than Murban because of the higher compositions in organic acids, sulfur and mercaptans while Murban was ahead only regarding the content of salts.

Chemical Compositions of Metals

The elemental compositions of used metals according to the XRF detector are given in the Table 3.

Table 3: The elemental chemical compositions of used metals

Metal	Fe (%)	Mn (%)	Co (%)	Ni (%)	Cr (%)	Cu (%)	P (%)	Mo (%)	Si (%)	S (%)	Ti (%)	V (%)
(1)Carbon Steel (High)	98.60	0.43	-	0.17	0.14	0.37	0.12	0.086	0.09	-	-	-
(2)Carbon Steel (Medium)	99.36	0.39	-	-	-	-	0.109	-	0.14	<0.02	<0.04	-
(3) Carbon Steel (Mild Steel)	99.46	0.54	<0.30	-	<0.07	-	-	-	-	-	<0.19	<0.07
(4) 410-MN: 1.8 420-MN: 2.8 (Stainless Steel)	88.25	0.28	-	0.18	10.92	0.10	0.16	-	0.11	-	-	-
(5) 410-MN: 1.7 420-MN: 1.7 (Stainless Steel)	87.44	0.30	-	-	11.99	-	0.18	-	0.09	-	-	-
(6) 321-MN:1.4 304-MN:1.9 (Stainless Steel)	72.47	1.44	-	8.65	17.14	-	0.18	-	0.12	-	-	-
(7)Monel 400	1.40	0.84	0.11	64.36	<0.04	33.29	-	-	-	-	-	-

According to the obtained results for the chemical compositions of ferrous metals there were found some higher amount of ferrous in three types of carbon steels, moderate amount of ferrous in three types of stainless steel and trace amount of ferrous in Monel although stainless steels were composed with some other trace elements such as Mn, Co, Ni and Cr based on the enhancements of some specific properties [1] [3].

Average corrosion Rates of Metals

The average corrosion rates of metals with respect to both crude oils are given in the Figure 2.

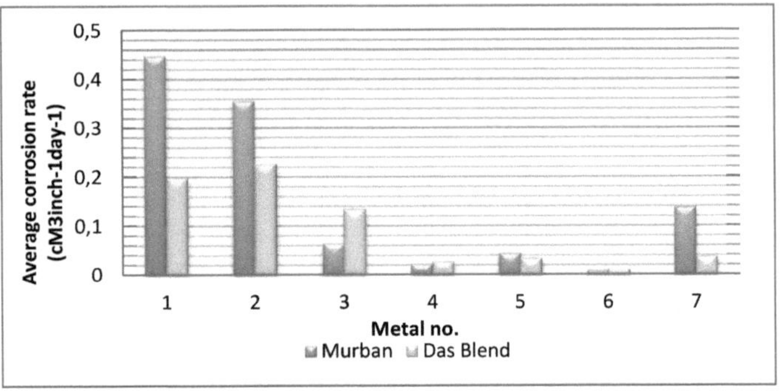

Figure 2: Average corrosion rates of metals in both crude oils

According to the obtained results three types of carbon steels showed relatively higher rates of corrosion than the other metals regarding both crude oils. The least corrosion rates were found from stainless steels while Monel was showing moderate corrosion rates in both crude oils. In accordance the chemical composition of seven different types of ferrous metals stainless steel was composed with moderate amount of ferrous while containing some significant amount of chromium and nickel which tend to form a corrosive protection film on the relevant metal surface itself in the wake of at least 12% of chromium is sufficient for the corrosive protection film [1] [4] [5]. In the 321-MN:1.4304-MN:1.9(Stainless Steel) which showed the least corrosion rate in both crude oils the chromium amount was 18%with higher amount of nickel and also other two types of stainless steels showed lower corrosion rates in both crude oil which were composed 12% and 13% of chromium. When considering the corrosion rates of metals accordance of the properties of both crude oils there were found some higher rates of corrosion of four types of metals in Murban crude oil than the Das Blend crude oil although greater amount of organic acids, sulfur and sulfur compounds were composed in Das Blend crude oil than the Murban crude oil so that there can be concluded the effect of salts in the metallic corrosion was stronger than the overall effect of mercaptans, sulfur and organic acids in the metallic corrosion. Because there cannot be expected a proper progress of sulfidation process due to the requirement of high temperature [8].

Qualitative Analysis of the Corrosion

According to the microscopic analysis of the surfaces of metal coupons some important characteristics of corrosion compounds were identified based on their visible appearances. A few of them are given in the Figure 3.

(a)

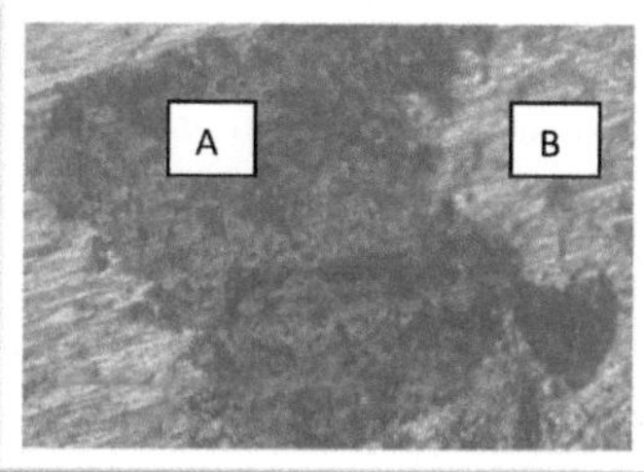

(b)

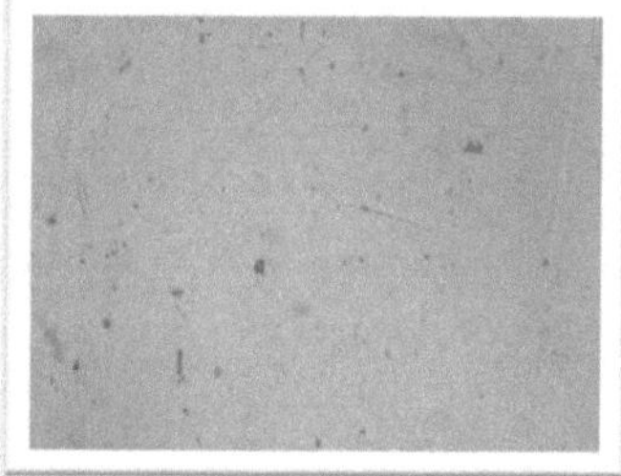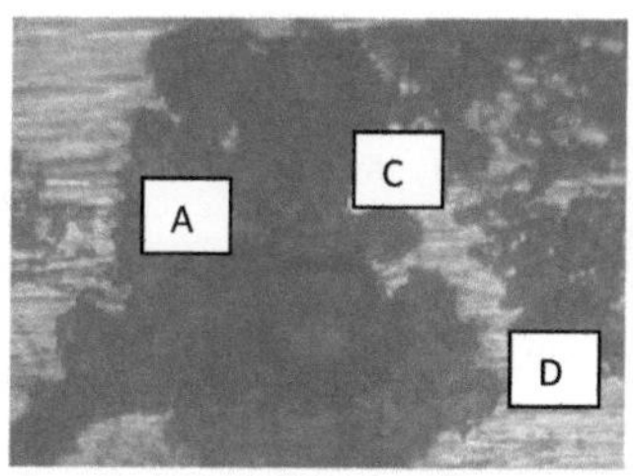

(c)

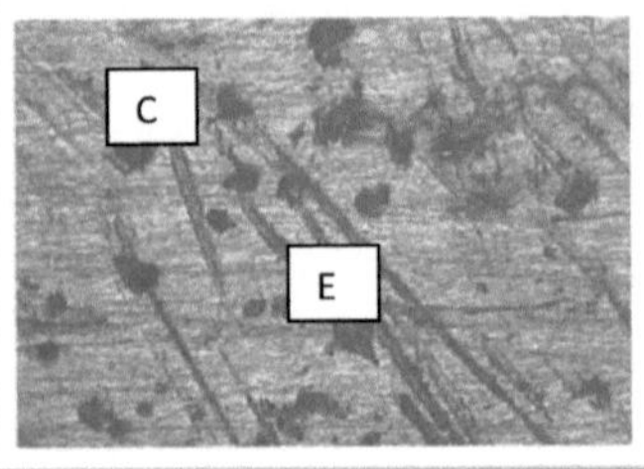

Figure 3: Corroded surface of (a) High carbon steel in Murban (b) Stainless steel in Murban (c) Monel in Murban

When considering the observed figures under the microscopic analysis there were identified some important and essential visible features regarding the cause of corrosion. The identifications were based on the physical appearances foremost the color [3]. A brief summary about the appearances and resent observations regarding the current research are given in the Table 4.

Table 4: Corrosion compounds and their appearances

Compound	Appearances	Observations
FeS	Black, brownish black, property of powder, pitting, cracks	Observed most of features in each metal piece.
Fe_2O_3	Rusty color	Observed rarely.
CuS	Dark indigo/ dark blue	Unable to specify.

According to the observed results of the microscopic analysis that can be concluded the major corrosion compounds and some specific features that have been formed on the metal surfaces as listed in below.

- ✓ A- Ferrous Sulfide
- ✓ B- Pitting Corrosion
- ✓ C- Corrosion Cracks
- ✓ D- Ferrous Oxides and Trace Compounds
- ✓ E- Copper Sulfide and Trace Compounds

Beside of the formation of corrosive compounds there were identified corrosion cracks in most of times on the metal surfaces especially in stainless steels and cavities rarely on the Monel metal piece. Those features can be used as confirmation stages of the metallic corrosion.

Decay of the Metals in Crude Oils

According to the results of atomic absorption spectroscopy (AAS) the ferrous concentrations and copper concentrations in crude oil samples are given in the Figure 4.

(a)

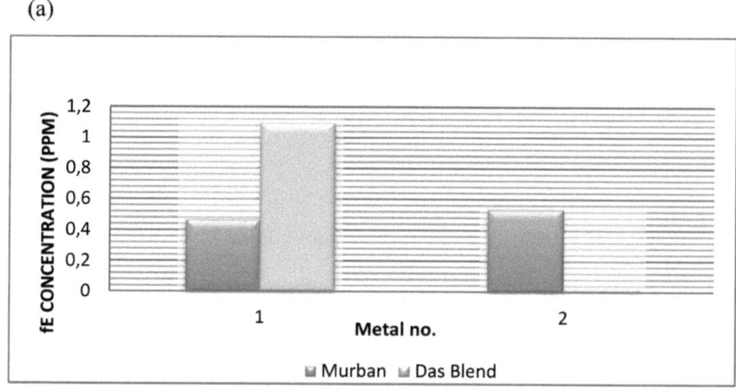

(b)

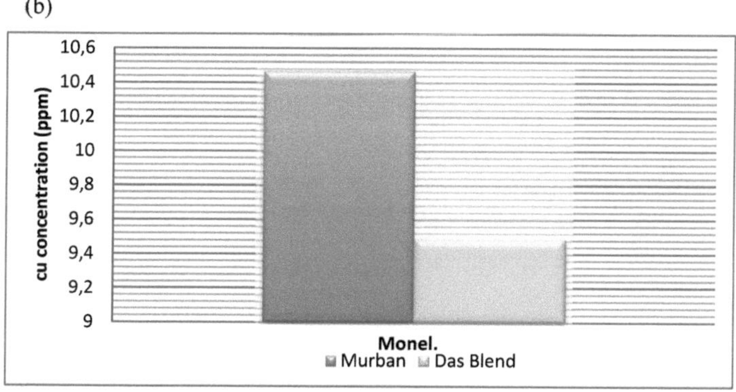

Figure 4: (a) Fe concentrations and (b) Cu concentrations in crude oil samples

According to the atomic absorption spectroscopic (AAS) results there were found relatively higher ferrous concentrations in both Murban and Das Blend crude oil samples which were exposed to high carbon steel and medium carbon steel also there were not found any amount of ferrous in both Murban and Das Blend crude oil samples which were exposed to mild steel or any stainless steel while finding significant concentrations of copper in both crude oil samples which were exposed to Monel. Regarding the formation of corrosion on the surfaces of ferrous metals usually formed the metal oxides, sulfides or hydroxides on the metals surface. Those corrosion compounds tend to remove from the metal surface into the surrounding systems due to the repulsive and attractive forces in between successive electrons and protons [3]. That indicated the reason for the observed invisible weight loss of metal coupons while determining the corrosion rates of metals also can be confirmed the formation of the metallic corrosion.

Variation of the Hardness

The variations of the hardness of each metal coupon with respect to both crude oils are given in the Figure 5.

(a)

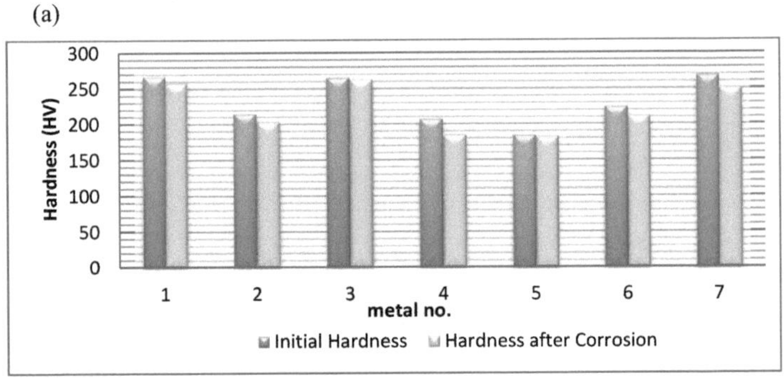

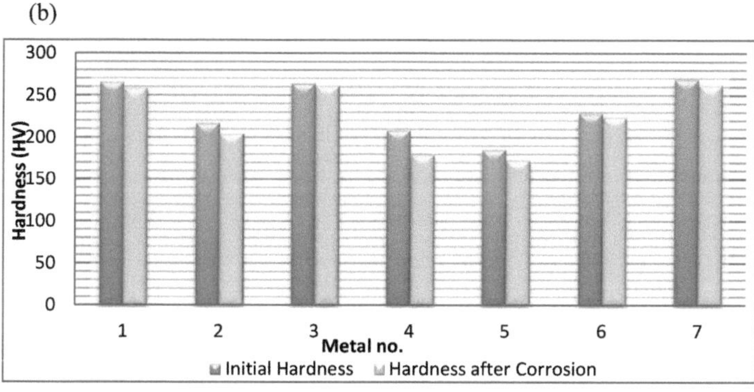

Figure 5: Variations of hardness of metals in (a) Murban and (b) Das Blend

The variations of hardness showed the slight reduction in each metal piece with respect to the both crude oils althoughby referring the distribution of hardness there cannot be identified any specific sequence or unable to express any linkage with the rate of the corrosion in the particular metal. The hardness was measured according to the point on the metal surface randomly. Therefore it might not distributed symmetrically. As a results of the formation of corrosive compounds on the metal surface those compounds tend to remove from the relevant metal surfaces while causing some unstable conditions on the metal surfaces also there were observed some asymmetric distribution of corrosion on the metal surfaces under the microscopic analysis [1] [2]. There can be concluded the reductionsof hardness in metals were happened due to the instability of corroded surface and the asymmetric distribution of the corrosion.

Conclusion

According to the obtained results there were found the cause of corrosion in each type of metal. The least corrosion rates were found from 321-MN:1.4304-MN: 1.9 (Stainless Steel) with respect to both crude oils which has a chemical composition of 18% chromium and relatively higher corrosion rates were found from three different types of carbon steels while some intermediate corrosion rates were finding from Monel in both crude oils. There were observed some asymmetric distribution of corrosion on each metal piece under the microscopic analysis while observed FeS, corrosion cracks, pitting corrosion and Fe_2O_3 regarding most of ferrous metals according to their visible appearances foremost the color although there were observed same features on the surfaces of Monel metal most similar to CuS. Some higher concentrations of ferrous were found form crude oil samples which were interacted with high carbon steels and medium carbon steels while there were not finding any amount of ferrous in crude oil samples which were interacted with any stainless steel although found relatively higher concentrations of copper in both crude oil samples that exposed to Monel. There were found slight reduction of hardness in each metal coupon with respect to both Murban and Das Blend crude oils. Finally those observations can be used as the confirmation evidences of the cause of metallic corrosion due to the effect of corrosive properties of both Murban and Das Blend crude oils.

Acknowledgement

The great assistances of the laboratory staff at the refinery of Ceylon Petroleum Cooperation, laboratory staff at the Uva Wellassa University and laboratory staff at the University of Moratuwa will be appreciated prestigiously.

References

[1]. Khana O.P., Materials Science and Metallurgy, New Delhi: DhanpetRai and Sons publication, 2009.

[2]. Fahim M.A, Alsahhaf T.A and Elkilani A., Fundamentals of Petroleum Refining, Amsterdam: Radarweg Press, 2010.

[3]. Calister W. D., An Introduction of Materials Science and Engineering, New York: John Wiley & Sons, Inc, 2003.

[4]. Davis M.E. and Davis R.J. (1st Ed.), Fundamentals of Chemical Reaction Engineering, New York: McGraw-Hill, 2003.

[5]. Singh R., Introduction to Basic Manufacturing Process and Engineering Workshop, New Delhi: New Age International Publication, 2006.

[6]. Bolton, W. (2nd Ed.), Engineering Materials Technology, London: B. H Newnes Limited, 1994.

[7]. Badmos A. Y., Ajimotokan H. A. and Emmanuel E. O., Corrosion in Petroleum Pipelines, 2009, 36-40.

[8]. Speight, J.G. (3rd Ed.), The Chemistry and Technology of Petroleum, New York: Marcel Dekker, 1999.

[9]. Afaf G. A., PhD. Thesis, University of Khartoum, 2007.

[10]. Oparaodu, K. O., Okpokwasili, G. C., Comparison of Percentage Weight Loss and Corrosion Rate Trends in Different Metal Coupons from two Soil Environments, 2014, 243-249